W9-AWX-927

Better Homes and Gardens®

BIRD BUDDIES

Hi! My name is Max.
I have some great projects
to show you – and they're all
about birds! We're going
to have lots of fun making
them together.

C.1

©Copyright 1989 by Meredith Corporation, Des Moines, Iowa.
All Rights Reserved. Printed in the United States of America.
First Edition. First Printing
ISBN: 0-696-01892-6
MAX THE DRAGON™ and other characters in this book are trademarks and copyrighted
characters of Meredith Corporation, and their use by others is strictly prohibited.

Inside You'll Find...

Learn to identify tropical birds and count them, too.

Beautiful Birds

Max and his best friend, Elliot, are
spending the afternoon bird-watching.
There are many different birds to see!

Bird Names

Did you know that different birds have different names, just like you and your friends do? Look at the birds in the picture. How many of the birds' names do you know? Can you match the birds below with the birds in the big picture? Count them, too!

1  Peacock (PEE kok)

2  Toucans (TOO kanz)

3  Cockatoos (KOK a tooz)

4  Parrots (PARE uts)

5  Flamingos (fla MIN goz)

Colorful birds with paper-plate tails and fuzzy feet.

Rainbow Birds

The colors and the shape of a rainbow look like the tails of these Rainbow Birds. Would you like to make a Rainbow Bird?

What you'll need...

- Newspapers or brown kraft paper (optional)
- Scissors
- 1 paper plate
- Paints or colors
- Paintbrush
- Glitter (optional)
- Stickers (optional)
- White crafts glue
- Bird Head and Body (see page 30)
- One 6-inch piece of pipe cleaner
- Tape

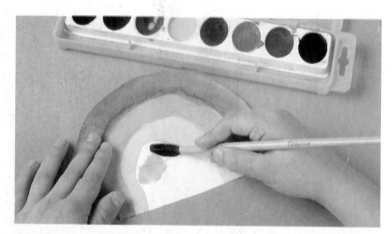

1 If desired, cover your work surface with newspapers. To make the bird tail, cut a 6-inch circle out of a paper plate or use a 6-inch paper plate. Cut a narrow strip off the bottom of the paper circle. Paint the paper circle (see photo). If desired, decorate with glitter or stickers.

2 Glue the Bird Head to the Bird Body. Glue the Bird Head and Body to the bird tail (see photo). Make a face any way you like on the Bird Head.

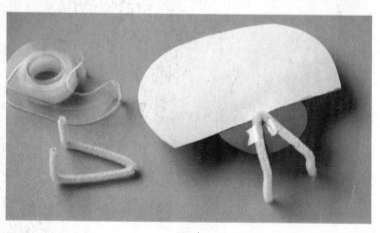

3 For the legs and feet, fold the pipe cleaner in half. Bend each end of the pipe cleaner out about ¾ inch. Tape the feet to the back of the Bird Body (see photo). Adjust the feet so your Rainbow Bird will stand up.

Create spectacular birds with baked homemade clay.

Pretty Birds

Making these Pretty Birds is like making cutout cookies except you don't eat them. And you use paint instead of frosting.

What you'll need...

- Tape
- Waxed paper
- Flour
- Salt Dough (see page 30)
- Rolling pin
- Cookie cutters
- Plastic drinking straw
- Shortening
- Baking sheet
- Paints
- Paintbrush
- Ribbon or fabric (see tip on page 9)

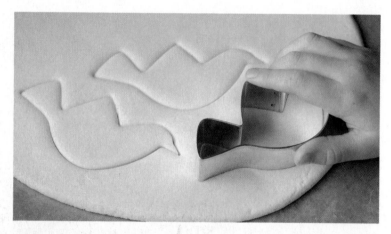

1 Tape a piece of waxed paper to the counter. Sprinkle some flour on it. (This keeps the dough from sticking.) With your hands, flatten the dough slightly. Roll or pat the dough to ¼-inch thickness. Cut bird shapes from the dough using cookie cutters (see photo). Carefully remove the extra dough around the bird shapes.

2 Using one end of the drinking straw, poke a hole near the top of each bird for hanging (see photo). Then poke one or two holes in the middle of each bird for the wing feathers. If desired, poke another hole for tail feathers (see photo on page 9). Carefully lift birds and place them on a greased baking sheet.

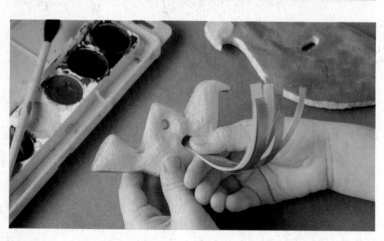

3 With adult help, bake in a 325° oven for 40 to 50 minutes or till dry. Cool completely. Decorate with paints. Let dry. Thread the ribbon through the hole in the bird body to make the wings (see photo). Repeat for the tail. If desired, add yarn for hanging.

Bird Feathers

Any colorful scraps of fabric or ribbon will work to decorate your bird's wings and tail. For our larger birds, we used strips about 1½ inches wide and 9 inches long for the wings, and 18 inches long for the tail. But you can cut them any length you like.

9

Search in the picture for the hidden ducklings.

Baby Birds

Oh, look what Max found! Ducklings. Two of the baby ducks are with their mom, but 7 of their brothers and sisters have wandered off. Can you help Max find the rest of the ducklings?

Did you know...

● Ducklings and all other birds hatch from eggs. Most baby birds stay in the nest for several weeks or months after hatching. Their parents feed and protect them until they can take care of themselves. Almost all birds leave their parents when they're a few months old to start their own lives in the wild.

● Look at the four pictures below. A duckling or baby duck is hatching from its shell. Do you know which picture comes first? Second? Third? Last?

Fruit or jelly beans are the eggs in this nest.

Nests to Nibble

Max just loves to eat fruit. And this fruit-filled bird nest is one of his favorite snacks. Sometimes he eats it with yogurt on top and sometimes without. He thinks it's yummy either way!

What you'll need...

- Mixing bowl
- 2 large shredded wheat biscuits
- Measuring cups
- Measuring spoons
- ¼ cup coconut
- 1 tablespoon brown sugar
- ¼ cup margarine or butter, melted
- Muffin tin
- Foil
- Fruity Bird Eggs (see tip on page 13) or jelly beans

1 To make the nests, in a mixing bowl crumble shredded wheat biscuits with your fingers (see photo). Use a spoon to stir in coconut and sugar. With adult help, pour in the melted margarine. Stir everything together.

2 With adult help, line each of the 6 muffin cups with a piece of foil. Press the shredded wheat mixture onto the bottoms and up the sides of the foil-lined cups (see photo). With adult help, bake in a 350° oven about 10 minutes or till crisp. Cool the nests in the cups.

3 Remove the nests from cups by lifting up on the foil. Carefully peel the foil off nests. Fill the nests with Fruity Bird Eggs. If desired, top the fruit with a spoonful of yogurt.

12

Fruity Bird Eggs

Fill your bird nest with different kinds of fruit, or pick just one. Try red or green grapes, blueberries, strawberries, melon balls, or drained canned fruit.

Our cotton-ball baby birds feel soft and fluffy.

Wee Wobblers

Look what just hatched! Our baby chicks are still a little wobbly. Give them a gentle push and watch what happens.

What you'll need...

- White crafts glue
- 2 cotton balls (any color)
- Scissors
- 2 tiny construction-paper circles, buttons, or shaky eyes
- Construction paper
- Clay or play dough
- Plastic eggshell halves or clean, fresh eggshell halves

1 For the baby chick's body, glue the cotton balls together (see photo). Glue on the paper circles for the eyes. Cut out a bird's beak from construction paper and glue it onto the cotton-ball face. (See tip on page 15.)

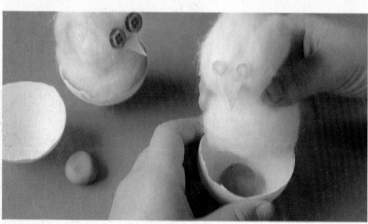

2 Roll a piece of clay into a ball about the size of a marble. Press the clay into the rounded bottom of the plastic eggshell half (see photo). Place your baby chick into the eggshell half.

Gently tap the egg and watch it wobble around.

Wobbler Faces

Eyes: If you have a paper punch at home, try using it to make eyes for your wobbling pals. You can punch holes out of paper, ribbon, or fabric. Then glue them on the bird's head.

Beak: Fold a small piece of paper in half. Then cut a tiny triangle on the fold. Glue it on the bird's head.

Birds in Flight

Max is visiting a flying school. His bird friends are learning how to fly. While the birds study about flying, can you help Max find the letters A, M, O, S, and T hidden in the picture?

Did you know...

● The first airplanes that were built couldn't fly very well. But they fly well now. That's because the airplane inventors decided to build airplane wings that work like bird wings.

● All birds have wings. Some birds can fly *very, very* fast. Much faster than you can ride in a car. No other animals can travel faster than birds.

● Some birds can't fly at all even though they have wings. Ostriches have long legs for walking or running on land. They use their wings for balance like you use your arms when you try to walk in a very straight line. Penguins can't fly either. They walk or use their wings like flippers so they can swim.

ostrich   penguin

Fashion a paper plate into a snazzy-looking bird face.

Best Bird Hats

Invite a flock of your friends over for a good time making bird hats. What kind of a bird is your hat? How can you pretend to be a bird? What sounds do birds make?

What you'll need...

- Scissors
- One 6-inch paper plate
- Crayons, colored pencils, or markers
- Construction paper
- White crafts glue or tape
- Stapler
- Pencil or a paper punch
- Yarn or string, cut about 12 inches long

1 Cut away up to half of the paper plate. Make a bird face on the paper plate by drawing with crayons. Or, cut shapes out of construction paper. Glue the shapes to the paper plate.

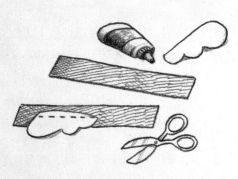

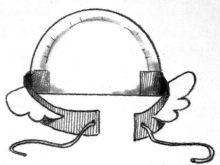

2 For the beak, fold a piece of construction paper in half. Cut a triangle on the fold of the paper. Glue the folded edge of the beak to the front of the paper plate.

3 For the headband, cut two 1½x6-inch strips of construction paper. If desired, cut 2 wings out of construction paper. Glue 1 wing to each headband strip.

4 Staple headband strips to both sides of the back of the plate. With adult help, use a pencil to poke a hole near the end of each headband strip. Thread yarn through holes.

Create a mobile with four strips of paper for each bird.

Whirling Paper Birds

What's light as a feather and flies in the breeze? Whirling Paper Birds! Hang them near an open window or blow on them gently so they'll "fly."

What you'll need...

- Scissors
- Four 1½-inch-wide strips of paper (see Step 1)
- Pencil
- Markers or paper
- White crafts glue or tape
- Yarn or string
- 1 drinking straw

1 Cut the strips of construction paper about 11 inches, 9 inches, 8 inches, and 4 inches long. For the tail, fold the 4-inch-long strip of paper in half lengthwise. Draw a triangle from the bottom corner to the opposite top corner. Cut out the triangle (see photo). For a feathery tail, make several cuts close together just to the fold.

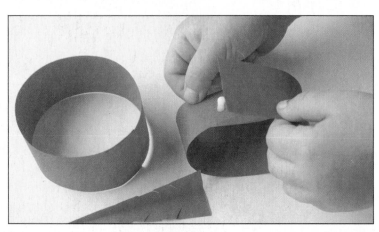

2 For the bird head, make a face any way you like with markers in the middle of the 9-inch-long strip of paper. Glue the ends of this strip together (see photo). For the bird body, glue the ends of the 11-inch-long strip of paper together.

3 For feathery wings, make several cuts close together on each end of the 8-inch-long strip of paper. Glue wings to the top of the body. Glue the wide end of tail to the bottom of the body. Glue the head to the wings (see photo). Tie a piece of yarn to the bird's head. Tie the other end to the straw. Repeat all steps to make a second bird.

A plastic-foam plate and buttons get this bird off the ground.

Soaring Bird

You're the power behind your Soaring Bird. Give it a gentle throw and watch it glide in for a terrific landing. Look, there goes Max's bird now! Up, up, and away!

What you'll need...

- White crafts glue
- Two ½-inch buttons
- Bird Glider (see page 32)
- Tissue paper or stickers

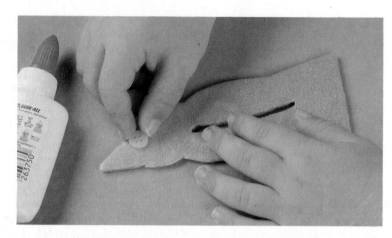

1 Glue the buttons to the Bird Glider's head for eyes (see photo). Let dry. (It's important to put the buttons where we have shown you so your bird will fly straight.)

2 Push the wings through the slot in the Bird Glider's body (see photo). You can decorate your bird by gluing on small pieces of tissue paper. Or, use stickers.

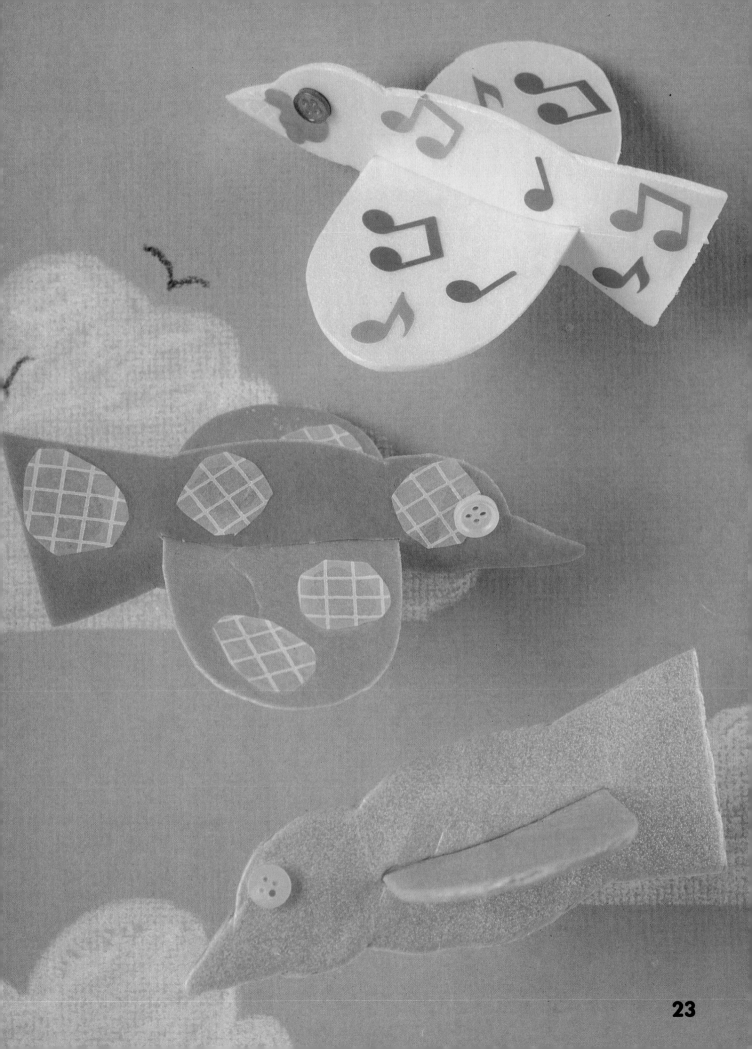

A rhyming word game.

Time to Eat

Max's friends are having some lunch.
And he's waiting to play with the whole bird bunch.
To help our dragon pass the time,
Look for all the words that rhyme.

Time for a Rhyme

Look inside the birdhouse for a word that rhymes with each of the words below. (You'll find the answers on page 32.)

sock

bear

moon

cat

kite

tower

Little snackers will enjoy fixing this no-mess treat.

Shake-a-Snack

You can put this snack together in a snap. Just put everything in a bag and shake, shake, shake!

What you'll need...

- Measuring cups
- 4 cups popped popcorn
- 1 cup desired cereal
- ½ cup mixed dried fruit bits
- ¼ cup tiny marshmallows or chopped nuts
- Sealable plastic bag or paper sack

1 Put the popcorn, cereal, fruit, and marshmallows into the plastic bag (see photo).

2 Close the end of the bag. Shake the bag to mix well (see photo).

Snack Talk

You can make this snack taste different every time you mix it together. Just use different kinds of cereal or dried fruit. Or, try using two or three different kinds of cereal in one batch of Shake-a-Snack.

You need only three ingredients for these bird feeders.

Feathered-Friend Feeders

Help the birds find these snacks by hanging them in a tree. Then watch for them to gobble up the special treats you made.

What you'll need...

- Cookie cutters (optional)
- Stale bread
- Table knife
- Peanut butter
- Sunflower nuts or wheat germ
- Pie plate
- 1 plastic drinking straw
- Tape
- Yarn or string, cut about 12 inches long

1 If desired, use cookie cutters to cut out shapes from stale bread (see photo). Or, you can leave the bread whole.

2 Using a table knife, carefully spread peanut butter onto 1 side of the bread. Place the nuts into the pie plate. Lay the bread, peanut butter side down, in the nuts to coat it well (see photo).

3 Use a drinking straw to poke a hole near the center of the bread. Wrap a piece of tape around the end of a piece of yarn. Pull the yarn through the hole in the bread (see photo). Cut the tape off. Hang the bread in a tree for the birds to eat.

Parents' Pages

We've filled this special section with more activities, recipes, reading recommendations, hints we learned from our kid-testers, and many other helpful tips.

Pretty Birds

See pages 8 and 9

You don't need to have bird-shaped cookie cutters to do this project. Trace around a picture of a bird and cut it out. Use that pattern to cut out shapes from the dough.

Salt Dough

    3 cups all-purpose flour
    1 cup salt
    1 cup water

● Combine flour and salt. Add water. Mix thoroughly.
● Turn dough out onto a lightly floured surface. Knead about 10 minutes or till you form a medium-stiff smooth dough. Cover and store the dough in the refrigerator until you are ready to use it.
● Before using the dough, check it for consistency. If the dough is too sticky, add a small amount of flour. If the dough is too stiff, knead in several drops of water. (This dough is intended for crafts projects only. Please do not eat it.)

Baby Birds

See pages 10 and 11

If you and your family enjoy watching birds, you might like to entice birds to nest in your backyard. Let your children help by providing nesting materials for the birds. Use a plastic berry basket and fill it with light and flexible materials, such as yarn, string, thin strips of woven fabric, cotton balls that are pulled apart, strips of facial tissue, or lint from your dryer. Then hang the basket in a nearby tree.
● Reading suggestion:
Make Way for Ducklings
    by Robert McCloskey

Beautiful Birds

See pages 4 and 5

Bird-watching is one of America's most popular pastimes. And why not? It's easy, you can do it anytime, and it's inexpensive. Try it with your children. Begin by getting a book from your library with pictures of different types of birds. Pack a picnic lunch or snack, and if you own a pair of binoculars, pack it too. Then set out for an adventure. Help your children identify some of the birds in your area. Which are their favorites?

Rainbow Birds

See pages 6 and 7

Here's a handy tip. To make your bird stand steady, tape the bottom of the feet to a rock, stick, or piece of paper. Or, pin the bird on a bulletin board.

Bird Body and Head: On the construction paper, trace around the top and bottom of a 3-ounce paper or plastic cup. Cut out the circles and use the bigger circle (2½ inches) for the body and the smaller circle (1½ inches) for the head.

Jewelry Gems

You can use Salt Dough to create many projects. Try designing your own fun-to-wear jewelry. To make a necklace, put a hole near the top of each dough cutout. Or, push a paper clip into the top of the cutout. After the dough is baked and decorated, thread a ribbon or a piece of yarn through the hole and tie a knot at the end.

Nests to Nibble

See pages 12 and 13

Bake an egg in a potato nest for breakfast or dinner.

Nested Eggs

 2 large potatoes, peeled and quartered
 ¼ cup milk
 ½ cup shredded cheddar cheese (2 ounces)
 ¼ cup shredded carrot
 ¼ cup shredded zucchini
 ¼ teaspoon salt
 4 eggs

● Cook potatoes, covered, in boiling water for 20 to 25 minutes or till tender. Drain and mash potatoes.
● Add milk to mashed potatoes; mash till combined. Stir in cheese, carrot, zucchini, salt, and ⅛ teaspoon *pepper*.
● Spoon potato mixture into 4 lightly greased 10-ounce casseroles. With the back of a spoon, push potato mixture from center, building up sides. Carefully break one egg into the center of each casserole. Place casseroles on a baking sheet. Bake in a 425° oven about 15 minutes or till eggs are cooked to desired doneness. Makes 4 servings.

Wee Wobblers

See pages 14 and 15

Our kid-testers were all giggles when they worked on these bouncy baby chicks. They were fun because the children could make the chick special in their own unique way. You can make your children's art experience a positive one and encourage their creativity by offering them a variety of materials to work with. You might pick colored paper, fabric, ribbon, lace, buttons, dried beans, crafts feathers, or sequins. All are good additions to these cute cotton-ball critters.
● Reading suggestion:
Wings: A Tale of Two Chickens
 by James Marshall

Birds in Flight

See pages 16 and 17

By sorting the birds into sets or groups of similar objects, your children are learning to classify. This is an important step in their speech development. It also helps them understand that the same kind of things may be different in size, shape, and color.
● Reading suggestions:
Owl Moon
 by Jane Yolen
Dabble Duck
 by Anne Leo Ellis

Best Bird Hats

See pages 18 and 19

Here's a new idea for your next party. Have your partygoers make their own silly hats. Have two or three different examples made ahead for them to look at. Then put an assortment of supplies out on a table and turn them loose. Don't forget to take a picture of the kids in their stylish new hats.

Whirling Paper Birds

See pages 20 and 21

Our kid-testers had trouble using scissors to give their bird's feathers. Help your children improve their cutting skills by letting them practice often with blunt-tipped scissors.

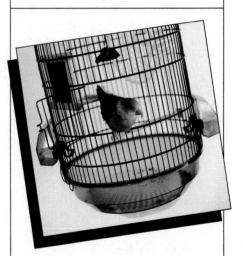

Birds as Pets

If you're thinking about buying a bird for a pet, be sure to find out as much as you can about the kind of bird you're considering. All birds have their own unique habits, and some birds naturally get along better with children than others. Also, if your children will help care for the bird, talk to them about the duties of a responsible, loving pet owner.

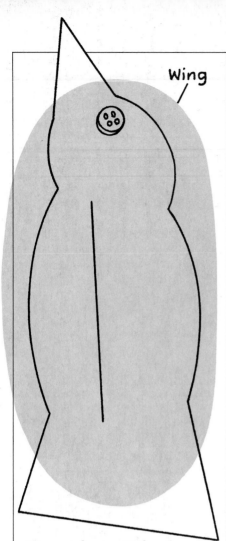

Wing

Soaring Bird

See pages 22 and 23

To ensure a good flight, we found that it's very important for the shape of the Bird Glider to match the pattern above.

Bird Glider: Trace the pattern onto a piece of paper and cut it out. Trace around the pattern pieces on a plastic-foam plate, meat tray, or egg carton. Cut it out. Use a sharp kitchen knife or crafts knife to cut the slit in the body.

Time to Eat

See pages 24 and 25

Answers to the activity on page 25: sock, clock; bear, chair; moon, spoon; cat, hat; kite, light; tower, flower.

Children have fun with rhyming words. And learning to recognize rhyming words makes them aware of the differences in sounds. You can help your children sharpen their skills by reading poems and rhyming books. Also, try improvising some of your own rhyming games. Invent the beginning of a couplet and have your children fill in the last word. "I wonder what you're going to do. I think you will put on your ————."

Shake-a-Snack

See pages 26 and 27

This snack is a good one to bag up for your hungry Halloween trick-or-treaters or to send to school, because it doesn't make little fingers sticky.

Feathered-Friend Feeders

See pages 28 and 29

Feeding the birds can be fun. If you and your children decide to start feeding the birds near your home, explain to them that once they start feeding the birds, they *must* continue. The birds become dependent on the food your children supply.

Feeding Fun

For the holidays, decorate an outdoor tree with food for the birds. Tie food on the tree with bright ribbons. Use food such as field corn, bread, suet, fruit, and garlands of popcorn and cranberries.

Remember, if you've been feeding the birds during the summer or fall, don't stop for the winter. It is *absolutely essential* to keep your feeders full of food during the winter.

BETTER HOMES AND GARDENS® BOOKS
Editor: Gerald M. Knox
Art Director: Ernest Shelton
Managing Editor: David A. Kirchner
Department Head, Food and Family Life: Sharyl Heiken

BIRD BUDDIES
Editors: Jennifer Darling and Sandra Granseth
Graphic Designers: Linda Ford Vermie and Brian Wignall
Editorial Project Manager: Angela K. Renkoski
Contributing Illustrator: Buck Jones
Contributing Photographer: Scott Little

Have BETTER HOMES AND GARDENS®
magazine delivered to your door.
For information write to:
ROBERT AUSTIN
P.O. BOX 4536
DES MOINES, IA 50336